CELL DIVISION & GENETICS

Robert Snedden

Series Editor
Andrew Solway

Heinemann Library
Chicago, Illinois

Designed by Paul Davies and Associates
Illustrations by Wooden Ark
Originated by Ambassador Litho Ltd.
Printed by Wing King Tong in Hong Kong

07 06 05 04 03
10 9 8 7 6 5 4 3 2 1

Library of Congress Cataloging-in-Publication Data
Snedden, Robert.
  Cell division and genetics / Robert Snedden.
      v. cm. -- (Cells and life)
Includes index.
Contents: The cell cycle -- DNA, the molecule of life -- Mitosis -- New
nucleus -- Cytokinesis -- Meiosis -- Life cycles -- Fertilization --
Development -- Animal development -- The gene -- The genetic code -- DNA
makes RNA makes protein -- Mutations -- Patterns of inheritance --
Genotype and phenotype -- Is it a boy or a girl? -- Variety, the spice
of life -- Smooth change, sharp differences -- Time for a change.
  ISBN 1-58810-672-1 (HC), 1-58810-934-8 (Pbk.)
  1. Cell division--Juvenile literature. 2. Genetics--Juvenile
literature. [1. Cells. 2. Genetics.] I. Title. II. Series.
  QH605 .S555 2002
  571.8'44--dc21
                                2001008580

Acknowledgments

The author and publishers are grateful to the following for permission to reproduce copyright material:
p. 4 P. Motta and Makabee/Science Photo Library; pp. 5, 22 Photodisc; p. 6 S. Stamers/Science Photo Library;
p. 7 N. Kedersha ULCA/Science Photo Library; pp. 8, 33b P. Motta and T. Naguro/Science Photo Library;
p. 10 A. Khodjakov/Science Photo Library; pp. 11, 32, 33a Eye of Science/Science Photo Library; p. 12
G, Schatten/Science Photo Library; p. 13 D. Scharf/Science Photo Library; p. 14 CNRI/Science Photo Library;
p. 16 A. Shay/Image Bank; p. 17 A. Ramage/Oxford Scientific Films; p. 18 Y. Nikas/Science Photo Library;
p. 20/Science Photo Library; p. 21 H. Reinhard/Oxford Scientific Films; p. 23 K. Lounatmaa/Science Photo
Library; p. 24 F. Whitehead/Oxford Scientific Films; p. 26 A. Pasieka/Science Photo Library; p. 29
D. Parker/Science Photo Library; p. 31 K. Edward/Science Photo Library; p. 35 R. Winslow, Animals,
Animals/Oxford Scientific Films; p. 36 Stone; p. 38 Biophotos Associates/Science Photo Library; p. 39 Hulton
Archive; p. 40 G.K. and Vikki Hart/ Image Bank; p. 42 J. Burgess/Science Photo Library; p. 43 D. J. Cox/Oxford
Scientific Films.

Cover photograph reproduced with permission of Science Photo Library/Addenbrookes Hospital.

Our thanks to Richard Fosbery for his comments in the preparation of this book, and also to Alexandra Clayton.

Every effort has been made to contact copyright holders of any material reproduced in this book.
Any omissions will be rectified in subsequent printings if notice is given to the publisher.

Some words are shown in bold, **like this.** You can find out what they mean by looking in the glossary.

Contents

1 Introduction

Why does a pig always give birth to piglets, not kittens? Why does a caterpillar, not a tadpole, hatch from a butterfly **egg?** These might seem like odd questions to ask, but they are at the heart of the study of **genetics.** Genetics is the study of heredity, or how the characteristics of living things are passed on from one generation to the next.

All living things are made of cells. Cells are life's building blocks. A single cell is too small to be seen without a microscope, but this tiny chemical package contains all the ingredients of life. Very simple forms of life are made up of a single cell. However, human beings, and all the plants and animals around us, are made up of millions upon millions of cells working together.

Dividing cells

Every living thing, from the smallest to the largest, begins life as a single cell. This single egg cell divides and divides many times, until it grows into an adult. Cell division is how **organisms** grow. When the organism is fully grown, some cells continue to divide to replace cells that grow old or are damaged in some way.

Each tiny egg cell holds the information it needs to grow into a complex organism, such as a cat or an oak tree. A sea urchin egg cell will grow into a sea urchin, while a human egg cell will become a new person. This information is contained in **DNA,** the cell's **genetic material.** DNA is an amazing substance that acts as an instruction book. It contains all the information a cell needs to reproduce itself.

This magnified photograph of a **eukaryote** cell shows the **nucleus** and other **organelles.** Magnification approx. x 23,000.

Animals such as this gibbon produce offspring that are similar to themselves. It passes on its characteristics to the next generation through the genetic material in its sex cells.

Reproduction

Bacteria and other microorganisms that consist of a single cell reproduce themselves by dividing in two. Most multicellular organisms cannot reproduce in this way. They are too complex. Instead, they produce special cells from which a new organism can grow. In sexual reproduction, cells from two parents combine to form a new living thing. For sexual reproduction, an organism has to make special sex cells. This involves a different kind of cell division.

In this book, we will look at how cells divide and find out more about DNA. We will then look at sexual reproduction and the different kind of cell division that this involves. Finally, we look at ways in which characteristics of living things are passed from one generation to another in their genetic material.

Prokaryotes and eukaryotes

There are basically two types of cells in the living world. Bacteria cells, the simplest living things, have no obvious structure. In particular, the DNA in a bacterial cell is not contained in a distinct nucleus.

The cells of all other living things are more complex. The inside of the cell is divided into various small compartments called organelles. The organelles keep different reactions in the cell separate and make it possible to gather all the materials for a particular job in one small area. The largest organelle in a cell is the nucleus, which contains the cell's DNA.

Bacterial cells are called **prokaryotes,** a word that means "before the nucleus," because they have no nucleus. The cells of other living things are called eukaryotes, which means "true nucleus."

2 The Cell Cycle

The series of events that take place in a cell from the time it divides until another division takes place is called the **cell cycle.** For a cell to divide, first it needs to make a copy of its essential parts. The most important part to be copied is the **genetic material** in the **nucleus.** This provides the instructions for the new cell to follow in maintaining, repairing, and copying itself. **Eukaryote** cells go through a series of changes as they copy the genetic material and then divide.

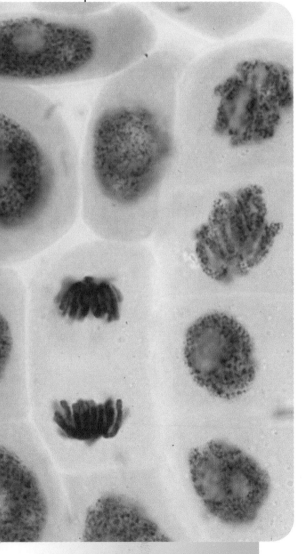

Cells in the growing root tip of a garlic plant (*Allium sativum*). The nucleus of the cell (lower center) has divided, but the cell itself has not yet split. Magnification approx. x 1,000.

The cell cycle begins when a new cell forms. It ends when the cell completes a division. How long this cycle takes is different from one cell type to another. Some cells divide more quickly. For example, cells that suffer a lot of wear and tear, such as those lining the small intestine in humans, might divide every eight hours or so. Roughly every five days, the lining of the small intestine is entirely replaced. Other cells divide more slowly. An average cell of a mammal divides about once every day.

The normal cell cycle of a plant or animal cell can be divided into two phases: a growth phase, called **interphase,** and a cell division phase. The cell division phase has two parts, **mitosis** and **cytokinesis.**

Interphase

Seen under a light microscope, a cell looks quiet during interphase. But in fact there is a great deal going on. **Proteins** and other materials in the cell are being produced, and the **DNA** in the cell's nucleus is being replicated, or copied. Also during interphase, the cell's **organelles** are replicated, so that each of the two new cells will have all it needs to begin work right away.

Interphase is divided into three stages called G1, S, and G2.
- During the G1 phase, the cell grows and prepares for DNA replication.
- In the S phase, a copy of the cell's DNA is made.
- In the G2 phase, the DNA has been replicated, and the cell prepares for division.

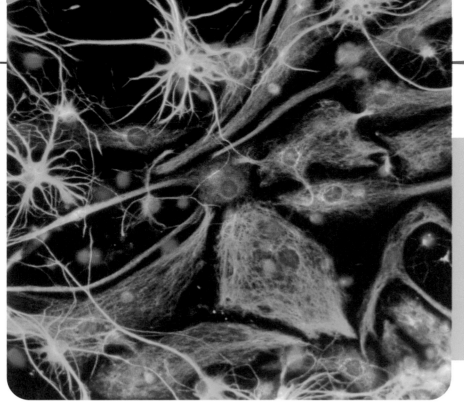

In most multicellular **organisms,** some types of cells become specialized and lose their capacity to divide. The star-shaped cells are specialized cells of this type. They supply nutrients to the nerve cells in the brain. Magnification approx. x 350.

The time a cell spends in G1 or G2 can vary considerably. Some cells, such as nerve and muscle cells, do not divide at all once they have become fully mature. They remain firmly in the G1 phase for months or even years, as they carry out tasks in the body. In actively dividing cells, interphase is a period of great activity. During G1 the cell can double in weight. This is when new organelles are formed.

When the cell reaches a certain size, it enters the S phase. If for some reason the cell cannot grow to the right size, for example, because there are not enough nutrients, then it does not divide. Instead, it enters a resting phase until conditions improve. However, once it reaches the critical size, it begins to copy its DNA and will continue through the other stages of cell division.

During G2, the cell structures needed for mitosis and cytokinesis are assembled. In a nondividing cell, the DNA molecules are like very thin threads. During G2, these DNA molecules begin to coil. Eventually the molecules form compact structures called **chromosomes.** The cell is now ready to begin mitosis.

Prokaryote cell division

The division of a **prokaryote** cell is a much simpler process than that in eukaryotes. First, the genetic material, a single DNA molecule, is replicated. Then both the original and the copy attach themselves to a different part of the cell membrane. When the cell begins to pull apart, the two DNA molecules are separated, forming two cells with the same genetic material. Because the process is much simpler, prokaryotes can divide more quickly than eukaryotes can. Under ideal conditions, some bacteria can divide once every 20 minutes.

New Nucleus

Every **eukaryote** cell has a **nucleus,** and every time a cell divides, a new nucleus is formed. In this way, **DNA** is passed on to the next generation of cells. **Mitosis** is the division of the cell nucleus into two identical copies. Each contains a complete set of **chromosomes,** which are a copy of the parent cell's DNA.

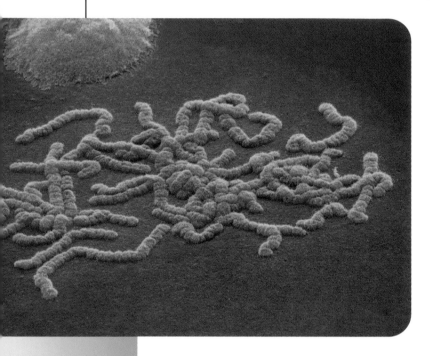

Mitosis begins after the cell has replicated its DNA and has finished constructing new cell parts. Mitosis can be divided into a number of stages, although there are no clear dividing lines between the stages. The four stages of mitosis are **prophase, metaphase, anaphase,** and **telophase.** The whole process, from prophase to telophase, usually takes between 30 minutes and three hours. Division of the cell itself, called **cytokinesis,** occurs after all four stages of mitosis end.

The nucleus of this cell is about to divide. The DNA has coiled itself into a number of compact chromosomes. Magnification approx. x 1,000.

Prophase

In **interphase,** the DNA in the nucleus is mixed with **protein** in a tangled mass called **chromatin.** During prophase, the DNA coils tightly upon itself, then coils even further around some of the proteins within the nucleus. This coiling causes the chromosomes to shorten and fatten, becoming thick, rodlike structures that are visible under the microscope. Each chromosome consists of a pair of identical structures called **chromatids,** joined in the middle at the **centromere.** This gives each pair of chromatids a rough X-shape.

When the chromosomes appear in the nucleus, the cell forms a structure called a **spindle.** The spindle is made from thin tubes of protein called **microtubules.** It separates the two sets of DNA. In the spindle of an animal cell, the microtubules are attached to two pairs of stubby, rodlike structures called **centrioles.** One pair of centrioles moves to each end of the cell, each one with a fan of microtubules radiating from it. At this point, the membrane surrounding the nucleus starts to break down. This event marks the transition from prophase to metaphase.

Metaphase

During metaphase, the chromatids move to the center, or equator, of the cell. The microtubules of the spindle become attached to the pairs of chromatids. Microtubules from both sides of the cell attach to the centromere of each chromatid pair. When the chromosomes are lined up at the cell's equator, anaphase begins.

Anaphase

At the beginning of anaphase, the microtubules of the spindle begin to pull apart the pairs of chromatids. One chromatid from each pair goes to one end of the cell, the other one goes to the other end. These will be the chromosomes of the two new cells. Anaphase lasts only a few minutes. By the time it ends, the spindle has disappeared and there are two identical sets of chromosomes, one at each end of the cell.

Telophase

During telophase, the new cell nuclei form. The chromosomes begin to unravel and become threadlike once again, disappearing from view when observed through a microscope. As this happens, new nuclear membranes begin to form around the chromosomes. Mitosis is now complete.

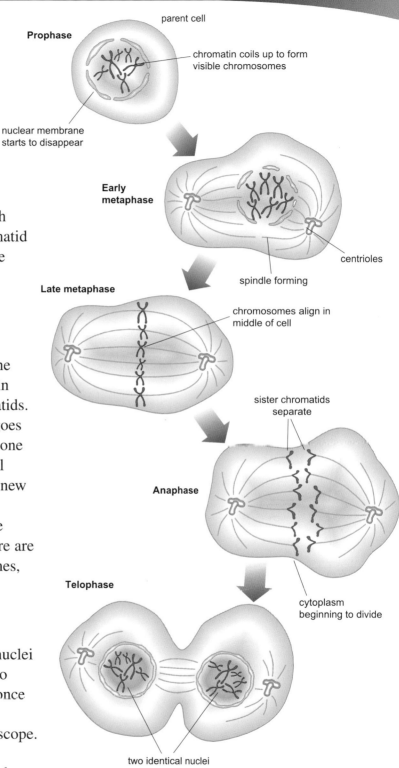

parent cell

Prophase

chromatin coils up to form visible chromosomes

nuclear membrane starts to disappear

Early metaphase

centrioles

spindle forming

Late metaphase

chromosomes align in middle of cell

sister chromatids separate

Anaphase

Telophase

cytoplasm beginning to divide

two identical nuclei

The four stages of mitosis are prophase, metaphase, anaphase, and telophase.

The Splitting of Cells

At some point between late **anaphase** and the end of **telophase,** the cell's **cytoplasm** divides to produce two separate daughter cells. This is called **cytokinesis.** Cytokinesis occurs differently in different **organisms.**

Animal cell division

Cell division in animal cells begins with a process called cleavage. All cells have a cytoskeleton, or cell skeleton. The cytoskeleton is a network of thin **protein** fibers called microfilaments. They are attached to the membrane and give the cell shape. Microfilaments have the ability to grow longer or shorter. At the beginning of cleavage, a band of microfilaments around the middle of the cell begins to pull in the cell membrane, producing a deepening furrow, or indentation, around the middle of the cell. This cleavage furrow continues to get deeper until it splits the cell in two.

Each new cell has a complete cell membrane and its cytoplasm contains the **mitochondria** and other **organelles** necessary for a successful life. Each cell has its own **nucleus** with a full set of **chromosomes** that contain all the instructions for growing and dividing.

Plant cell division

Plant cells differ from animal cells in that they have an outer cell wall in addition to the cell membrane. The cell wall is too rigid to be pinched together as in the cleavage of an animal cell. Instead, a new wall forms down the middle of the cell.

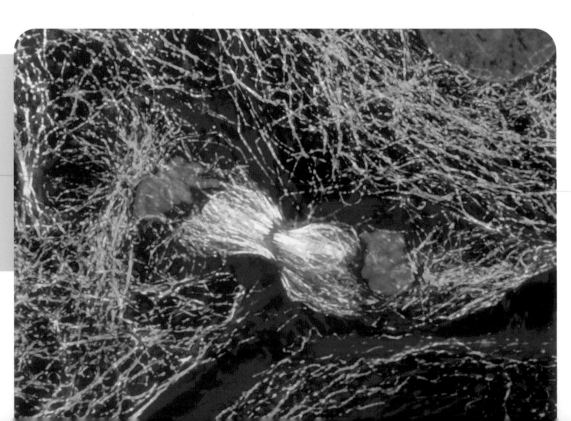

This electron micrograph of kidney cells shows cleavage about to take place. Magnification approx. x 1,000.

First, the cell produces small sacs called **vesicles,** which are stuffed with the materials necessary for building cell walls. These vesicles move to the equator of the cell, where they join with the remains of the **spindle.** The membranes of the vesicles begin to join to form two new membranes across the middle of the cell. At the same time, the building materials inside the vesicles form a disklike cell plate between the two membranes. The cell plate grows until it joins with the existing cell wall. The two new plant cell nuclei are now separated from each other by two membranes and cell walls.

Millions of divisions

It is the almost perfect repetition of the **cell cycle,** time after time, that makes us what we are. A single fertilized **egg** cell divides and divides until there are around ten trillion cells. As a result of this superbly accurate replication, every cell in the human body contains the same **genetic material** as the egg cell from which it grew.

This is the protistan *Plasmodium,* the microbe that causes malaria. At one point in its life cycle, *Plasmodium* forms nearly 1,000 offspring by multiple division. Magnification approx. x 3,400.

Multiple copies

Some of the simplest **eukaryote** organisms are single-celled creatures called **protistans.** A few protistans reproduce by multiple fission, in which the parent cell divides into a large number of tiny offspring. An example is the protistan *Plasmodium,* the parasite that causes malaria. When a type of malarial cell called a sporozoite infects a human liver cell, it divides by multiple fission to produce nearly 1,000 offspring cells. These are then released into the blood, where they attack red blood cells. This is what causes the fever associated with malaria.

3 Reproduction

The sequence of events that occurs as a living **organism** grows, develops, and eventually reproduces itself is called its life cycle. Growth and reproduction are always part of a complete life cycle.

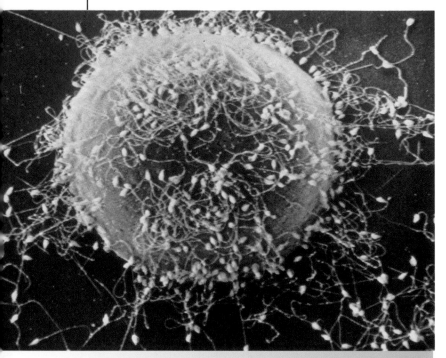

Sperm cells around a sea urchin egg cell. Only one of the sperm will penetrate the egg cell. Its nucleus will then fuse with that of the egg cell. Magnification approx. x 500.

Types of reproduction

Multicellular organisms can reproduce in one of two ways. Asexual reproduction is the equivalent of simple division in single-celled organisms. A single individual in some way produces one or more offspring. All the **genetic material** comes from the single parent, so parents and offspring are genetically identical, or clones.

Most multicellular organisms reproduce at least part of the time by sexual reproduction. In sexual reproduction, two individuals produce offspring that have genetic material from both parents. Sexual reproduction introduces new gene combinations in populations. Human beings and other living things are either male or female. To create a new human, a male sex cell, called a **sperm,** has to join with a female sex cell, called an **egg.** These sex cells, called **gametes,** are produced in special reproductive tissues or organs. The offspring is a unique individual and may have characteristics that are different from those of either parent.

Half the chromosomes

Every **species** has a certain number of **chromosomes** in its cells. These chromosomes exist in pairs. For example, human beings have 46 chromosomes in 23 pairs, while pea plants have 14 chromosomes in 7 pairs. The two chromosomes of a pair are not identical, but they are similar in size and shape. Most importantly, both chromosomes are responsible for producing the same characteristics in the cell. These pairs of chromosomes are called **homologous pairs,** meaning that each member of the pair is like the other. A cell with two of each type of chromosome is said to be **diploid.**

In sexual reproduction, the genetic material from two cells joins together. If both these cells had the normal number of chromosomes, then the new cell would have twice as many chromosomes. In the next generation, the chromosome number would double again, and so on.

Obviously the chromosome number cannot double in each new generation. To avoid this problem, when gametes form they have only half the usual number of chromosomes (one of each homologous pair). A cell with only one of each type of chromosome is said to be **haploid.**

Alternation of generations

In most plants, there are two distinct forms of the plant that occur alternately. This is called alternation of generations. One of the forms is diploid and is called the sporophyte generation. The sporophyte produces tiny haploid cells called spores. Spores are resting cells that can survive difficult conditions such as drought and extreme cold. Once conditions improve, spores grow and develop into the second form of the plant, called the gametophyte. The gametophyte produces male and female gametes, which join to produce a new generation of sporophytes.

Thus the life cycle of most plants alternates between producing spores and producing gametes. In simple plants such as mosses, the main form of the plant is the haploid gametophyte, with half the full set of chromosomes. The moss sporophyte is a tiny stalk and capsule that grows on top of the moss plant. In ferns, the parts that we see are the sporophyte plants, while the gametophytes are tiny. In flowering plants, which include most plants and trees, the main plant is the sporophyte. The male gametes are formed inside the pollen tubes that grow from pollen grains, while the female gametes are egg cells contained in structures called carpels. The gametophyte is not a separate plant but is hidden away in the flower.

This electron micrograph shows pollen grains from different flowering plants. Within each grain are haploid **nuclei.** The male gametes form from these nuclei after **pollination.** Magnification approx. x 300.

13

Meiosis

The **gametes** formed for sexual reproduction are **haploid.** That is, they have only half the usual number of **chromosomes.** When gametes are formed, a different kind of cell division takes place. In this type of cell division, the number of chromosomes in the new cells is halved. This special division of the **nucleus** is called **meiosis.**

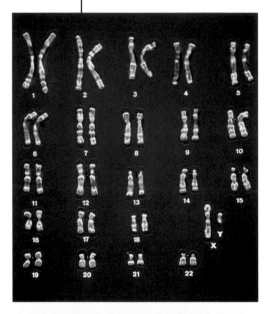

A photograph of the 23 pairs of human chromosomes. Magnification approx. x 2,800.

The chromosomes in a normal **eukaryote** cell are paired. The two chromosomes in the pair are not identical, but they do the same job. They are a **homologous pair.** Because of this pairing, it is possible for some cells with only half the chromosomes to function perfectly well. Indeed, plants such as mosses are haploid.

Gametes are formed from cells called germ cells, that are found in the reproductive organs. In meiosis, a germ cell divides not once but twice. Before meiosis begins, the germ cell's **DNA** is duplicated, just as it would be before **mitosis.** Exact copies of each chromosome are made, forming a joined pair of **chromatids.** Then the cell divides. The resulting cells then immediately divide again. However, the DNA is not replicated beforehand. The result is four haploid cells.

First division

During **prophase** of this first division (meiosis I), the duplicated chromosomes pair with their homologues. This joining of homologous pairs is unique to meiosis. It does not happen during mitosis. As in mitosis, the chromosomes coil and become visible under the microscope. The chromosomes pair together. Each chromosome forms from a pair of identical chromatids joined at the **centromere.** At this stage, another process unique to meiosis can happen. The homologous pairs of chromosomes may swap segments of DNA with each other. This is called crossing over. The crossing over process is an important way of producing a greater variety of characteristics in offspring. Following prophase is **metaphase,** when the chromosomes move to the center of the cell.

In **anaphase** of meiosis I, the homologous pairs of chromosomes separate and move to either end of the cell to form the new nuclei in **telophase.** Division continues and two new cells, called daughter cells, are produced. Each of these cells has only half the normal number of chromosomes. But each of the chromosomes has two chromatids, so there are two identical copies of the DNA.

Second division

In meiosis II, the second division, the two daughter nuclei formed in meiosis I divide to form four new nuclei. Meiosis II is similar to ordinary mitosis, but there is often no **interphase** before this division begins. In fact, there is no need for one, because there are already two copies of the DNA.

In meiosis II, the chromosome pairs move to the center of the daughter cells. As they do so, a **spindle** forms at either end of the cell and attaches to the chromosomes. The two chromatids in each chromosome are then pulled apart and move to opposite ends of the cell.

By the time meiosis II has been completed, there are four daughter nuclei in four separate cells. Each daughter cell has only half the number of chromosomes that the germ cells had, and it has only one copy of each of these chromosomes.

This diagram shows meiosis I, in which the nucleus divides and homologous pairs of chromosomes are separated, and meiosis II, when the nucleus divides again, separating the two copies of each chromosome.

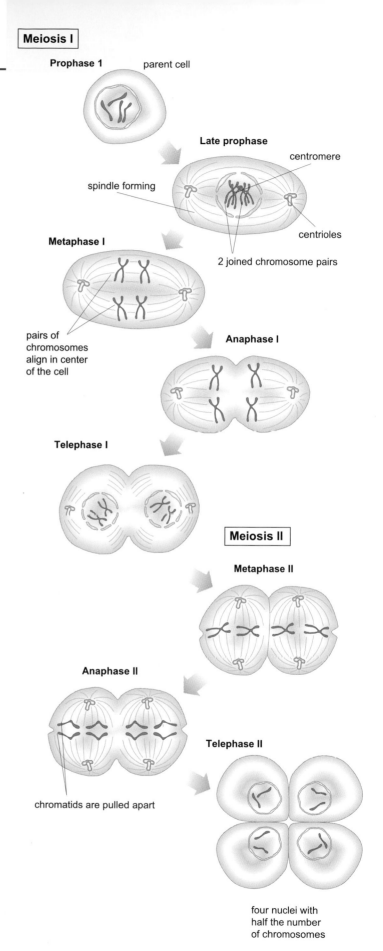

Meiosis I

Prophase 1   parent cell

Late prophase

centromere

spindle forming

centrioles

2 joined chromosome pairs

Metaphase I

pairs of chromosomes align in center of the cell

Anaphase I

Telephase I

Meiosis II

Metaphase II

Anaphase II

Telephase II

chromatids are pulled apart

four nuclei with half the number of chromosomes

Fertilization

In all living **organisms,** fertilization takes place when a male **gamete** and a female gamete join together. Both male and female gametes carry half the number of **chromosomes** (the **haploid** number) of the adult. When they fuse, a cell called a **zygote** is formed, which has the **diploid** number of chromosomes.

Bringing gametes together

The way fertilization takes place varies from **species** to species. In many plants, for example, fertilization can take place between gametes from the same individual (self-fertilization) or between gametes from different individuals (cross-fertilization).

Bees are probably the most important pollinators of many flowering plant species. Honey bees are capable of carrying large amounts of pollen in pollen sacs on their back legs.

Plant gametes

Primitive plants, such as mosses and ferns, require a moist environment so that male gametes can swim across the plant to reach **egg** cells and fertilize them. In flowering plants, the male gametes are carried in pollen grains to the **stigma,** the female part the flower. Pollen is most commonly transported to another plant of the same species either by insects or the wind. This transfer of pollen from the male to the female parts of flowers is called **pollination.**

Pollen grains that succeed in reaching a stigma begin to grow a microscopic pollen tube that penetrates down to the ovule. A **nucleus** from the pollen grain travels down the pollen tube to the ovule and fuses with the nucleus of the egg cell.

A female common frog (*Rana temporaria*) lays her eggs, or spawn. In most frog species, the eggs are fertilized externally, after they have been laid.

Animal gametes

In animals, fertilization is usually between different individuals. Fish and amphibians release their eggs and **sperm** into water, and the sperm swim to the eggs to fertilize them. This is called external fertilization. In land-living animals like the reptiles, birds, and mammals, the male inserts his sperm through an opening in the female's body. This is internal fertilization. The advantage of internal fertilization is that mating does not need to take place in a wet environment for the sperm to swim to the egg cells.

In male animals, the sperm are very small, are produced in huge numbers, and develop flexible tails that enable them to swim toward an egg. Different species produce different types of sperm cells. The sperm of an opossum appear to have two tails, for example, because pairs of sperm swim together for greater efficiency.

In contrast to sperm, eggs are very large, are produced in small numbers, and are unable to move. The egg has all of the cell material and structure it needs to get the new individual off to a good start if fertilization takes place.

The male sperm swims toward the female egg and penetrates its outer layer. It then releases its nucleus into the egg cell, and this joins the egg's nucleus in a single diploid nucleus. This is the first cell of a new organism, the zygote.

4 Development

When a male and female **gamete** fuse to form a fertilized **egg,** or **zygote,** this marks the beginning of a potential new **organism.** But this is only the first step in a long and complex journey from that single-celled zygote to a multicellular organism. The process is called development, and the instructions that tell the cell how to develop are held in the organism's **genetic material,** its **DNA.** Development involves the shaping of the organism and the specialization of groups of cells to carry out particular tasks.

Animal development

Development in animals begins when the zygote, or new **diploid** cell, starts to divide repeatedly. Once it begins dividing, the zygote is called an **embryo.** At this stage, the contents of the original egg cell are simply being divided among a growing number of smaller cells, each with its own **nucleus.** There is no cell growth, and the embryo stays about the same size.

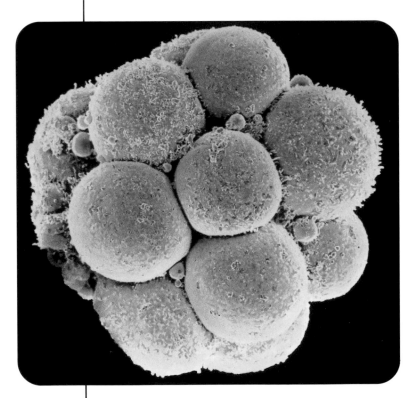

At first, the embryo consists of a ball of identical cells. This stage is called cleavage. Eventually a fluid-filled, hollow ball of cells called a **blastula** forms. The frantic pace of cell division slows down, and the embryo enters the next stage of development.

A human embryo four days after fertilization is still a ball of very similar cells. Soon these cells will begin to develop into different cell types. Magnification approx. x 900.

Germ layers

In the next stage of development, the blastula goes through a major reorganization. The cells of the blastula rearrange themselves to form a **gastrula** of two or three layers of cells, one inside the other. The cells that develop from these layers, called germ layers, will form all the tissues and organs in the adult organism.

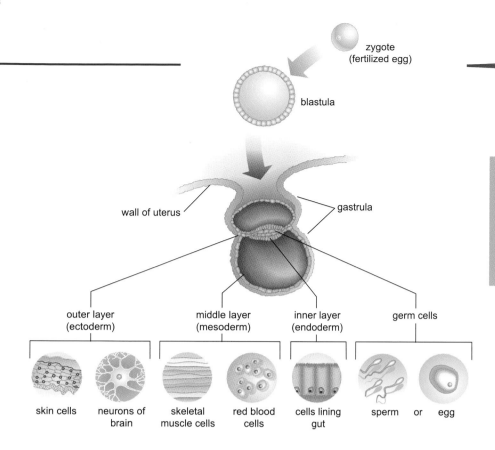

zygote
(fertilized egg)

blastula

wall of uterus

gastrula

outer layer
(ectoderm)

middle layer
(mesoderm)

inner layer
(endoderm)

germ cells

skin cells

neurons of
brain

skeletal
muscle cells

red blood
cells

cells lining
gut

sperm   or   egg

This illustration shows some of the different body tissues formed from the germ layers of the gastrula.

The outermost of the germ layers is the **ectoderm.** From this layer, all the cells of the skin and the nervous system develop. Within the ectoderm is the **mesoderm,** or middle layer. This layer produces the muscles and most of the skeleton, as well as the heart and blood vessels, the kidneys, and the reproductive organs. The innermost layer of cells, the **endoderm,** forms the lining of the gut and the organs associated with it.

Once the germ layers have formed, the cells in each layer quickly become distinct from one another as different groups of cells, or tissues, begin to specialize. This is the beginning of organ formation. Once the organs have formed, they grow and begin to function, doing the jobs they will perform throughout the life of the organism.

Guiding genes

One question studied by biologists is how a fertilized egg, one single cell, can develop into an organism with many millions of cells of many different types. The development process is directed and regulated by DNA in the cell nucleus. Sections of DNA, called genes, interact with one another to control development. Only a few genes concerned with the essential processes in the cell are active at all times. Other genes are turned on only at particular stages in development.

Growth and development

If an organism is going to grow bigger, its cells have to divide and make more cells. This is growth. Development, on the other hand, is the appearance of different, specialized body parts.

Plant Development

The following description of plant development relates mainly to **angiosperms,** the flowering plants. The vast majority of plants around us are angiosperms.

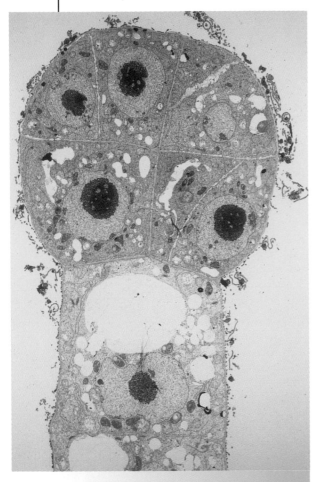

This dissected embryo of a turnip (*Brassica campestris*) shows the cell (bottom) that will develop into the first root and the cells (top) that will develop into the shoot and leaves. Magnification approx. x 500.

As soon as a plant cell is fertilized, the new **zygote** begins to divide. As in animals, it forms a group of cells, which then specialize to form an **embryo.** The embryo develops the beginnings of a young shoot, that will form the plant's stem and leaves, and a young root. The embryo also forms one or two seed leaves, or **cotyledons.**

The new embryo remains at the center of the ovule, the part of the flower where the **egg** cell originally formed. It is surrounded by a specialized tissue called the **endosperm,** that supplies the growing embryo with food. The ovule itself develops into a seed and forms a tough protective seed coat around the embryo and endosperm. While this happens, parts of the flower surrounding the seed develop into a fruit. As the fruit ripens, the seeds lose water and become dry.

Spreading the seeds

Once a seed is fully formed, the embryo inside stops developing, and the seed may become **dormant.** Before the plant can develop further, the seed must leave the parent plant.

Different plants spread their seeds in different ways. Some fruits are juicy, to attract animals to eat them. If they are eaten, the seeds survive the journey through the animal's gut and grow in its droppings. Some plants have very light seeds that are carried by the wind. In some cases, the seeds are attached to parachutes or wings that help them travel on the wind. Some fruits are pods that burst open and shoot the seeds out. There are also seeds that are carried by water, and seeds that hook onto passing animals.

Once a seed comes to rest, it will remain dormant until conditions are right to trigger the next stage in the plant's life. That stage is **germination.**

Germination

Germination is the restarting of growth of a plant embryo inside its seed after a period of dormancy. The most essential ingredient for germination to begin is water. The seed takes up water through a tiny hole in the seed coat. This causes it to swell. As a result, the seed coat bursts open. Once the seed has split, the young shoot begins to grow up toward the light, and the young root grows into the ground. The shoot is bent into a hook shape to protect its delicate growing tip as it pushes through the soil.

The cotyledons provide food for the early growth of the embryo. In plants such as broad beans, the cotyledons remain underground, inside the seed coat. Food stored in the cotyledons is converted into sugars, which can be transported to the shoot and root tips where growth is taking place. In other plants, the cotyledons are lifted up out of the soil and start to make food by **photosynthesis.** Once the new seedling grows its first green leaves, the cotyledons wither away.

Growing places

Plant growth is limited to specific growing zones called **meristems.** These areas of unspecialized cells are found at the tips of roots and shoots and as a layer of cells in the stems and roots. Plants are shaped through different patterns of growth at the meristems.

This photograph shows the germination of beech tree seedlings. The leaflike structures on the seedling on the right are the cotyledons.

Arrested development

The record for seed dormancy is held by the Arctic lupine. Seeds found in the frozen earth of the Yukon in Canada in 1954 were estimated to be between 10,000 and 15,000 years old. The seeds germinated successfully in 1966.

Time for a Change

Is there an internal aging clock in every **organism?** Fruit flies live for about a month, dogs for about fifteen years, humans perhaps more than 100 years if they are lucky. The consistency of the lifespan in individuals of a **species** suggests that genes are involved.

An organism ages throughout its lifespan. Changes in the structure of the organism and in its workings, especially after sexual maturity has been reached, affect every organ of the body. Eventually, these changes cause it to deteriorate physically and die. There are many theories about how and why aging occurs, but none yet accounts for all the known facts about the process.

Aging cells

In mammals, there are many changes as the body gets older. One of the main effects is a steady loss of muscle. However, as the muscles waste away, the amount of fluid and stored fat often increases so that overall the body becomes heavier. The body's **metabolic rate** slows, which reduces the amount of heat it produces. To make up for this, the body reduces the blood supply to the skin, in order to cut down on heat loss. The bones of the skeleton become weaker and are more easily broken. The skin becomes less elastic, which causes wrinkles.

Some body tissues are continually renewed by cell division. The cells of the liver and pancreas, for example, are constantly replaced by new cells, so the effects of aging are not so severe in these organs. Highly specialized tissues such as nervous tissue, however, do not divide once they are mature. These tissues show definite signs of deterioration as aging progresses. For instance, the number of nerve cells in the nervous system and brain falls steadily. Overall, cell division gets slower as the organism ages. Wounds take longer to heal and infections are harder to fight off because the **immune system** produces fewer cells.

As a person ages, changes in the skin make it less elastic, which leads to wrinkles.

Plants also age. Some plants seem to be genetically programmed to die after one or two years or even a few months. Although other kinds of plants can grow for years, their ability to do so declines with age.

Genes and aging

Studies of human cells cultured in the laboratory have revealed that a normal cell can usually divide only about 50 times before it dies. Most cells in the human body divide no more than 80 or 90 times. This limit is thought to be controlled by structures called **telomeres** on the tips of **chromosomes.** Each time a cell divides, part of the telomere on each chromosome is lost. When the telomere is lost completely, the cell dies.

Some people have suggested that the evolutionary value of aging lies in the act of giving the next generation a better chance to evolve and adapt to changes in the environment. This suggests that natural selection has favored genes that cause aging. Reproductive success in any organism means producing offspring and setting the next generation on its way.

Immortal cells

Some cells never lose their ability to divide. Examples are sex cells and cancerous cells. In these immortal cells, **enzymes** rebuild the telomeres after each cell division.

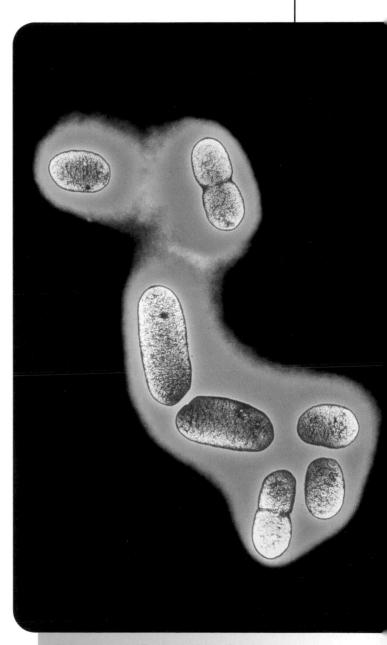

This bacteria (*Escherichia coli*) is in the process of dividing. Bacterial chromosomes are circular, so they do not have telomeres. As a result, they do not age as most **eukaryote** cells do. Magnification approx. x 10,000.

5 Genes and DNA

Each kind or **species** of **organism** has a set of characteristics that are passed on from generation to generation. A blackbird, for instance, will develop feathers and will be able to fly, while a beech tree will develop a trunk, bark, and leaves. However, individuals within each species are not exactly the same. There are some differences in their characteristics. A human may be tall or short, have blue eyes or brown eyes, or be male or female.

All these different inherited characteristics are carried by **DNA,** the cell's **genetic material.** DNA is the instruction manual that says this **egg** will develop into a human and not a halibut, or this seed will grow into an oak and not an orchid. It also influences whether a person will be tall or short, or fair-haired or dark.

There are many questions that we can ask about genetic material and how it works. What exactly are the instructions that are passed on from generation to generation? How do they produce their effects? Also, how are these instructions copied faithfully every time a cell divides?

By looking more closely at DNA, we can answer these questions.

The characteristics of these kittens are determined by the genes of their mother and father.

In the genes

The basic unit of inheritance, by which characteristics are passed from one generation to the next, is called the gene. When the word *gene* was first used at the beginning of the twentieth century, it referred to something passed from one generation to the next that determined a particular characteristic, such as the color of a person's eyes or the height of a pea plant. But at that time, no one was sure what a gene was.

A **chromosome** is a very long strand of DNA wrapped around **proteins.** We now know that a gene is one length of DNA. Genes are arranged along the chromosome like beads on a necklace. Each chromosome is likely to have thousands of genes. Within a species, the gene for a particular characteristic will always be found at the same place on the same chromosome.

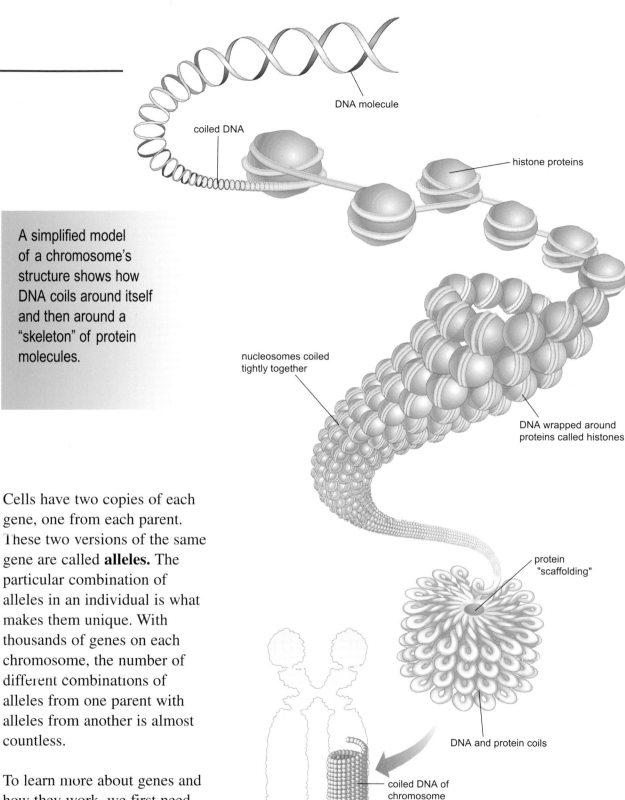

DNA molecule

coiled DNA

histone proteins

A simplified model of a chromosome's structure shows how DNA coils around itself and then around a "skeleton" of protein molecules.

nucleosomes coiled tightly together

DNA wrapped around proteins called histones

protein "scaffolding"

Cells have two copies of each gene, one from each parent. These two versions of the same gene are called **alleles.** The particular combination of alleles in an individual is what makes them unique. With thousands of genes on each chromosome, the number of different combinations of alleles from one parent with alleles from another is almost countless.

To learn more about genes and how they work, we first need to understand more about the structure of the DNA that chromosomes and genes are made of.

DNA and protein coils

coiled DNA of chromosome

DNA—the Molecule of Life

In the early 1950s, James Watson and Francis Crick, two scientists working at Cambridge University in Britain, figured out the structure of the **DNA** molecule for the first time. Crick was so thrilled by the discovery that he ran into a local pub announcing, "We've discovered the secret of life!"

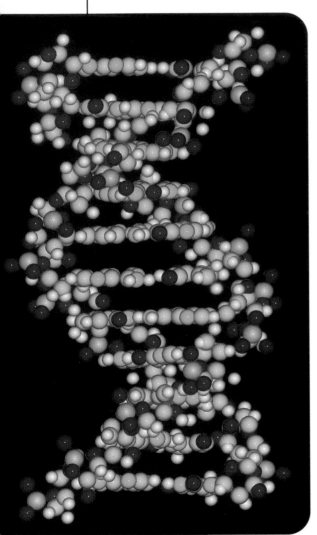

A computer model of part of a DNA molecule shows its twisted ladder shape. The colored spheres represent individual atoms.

More than any other substance, DNA defines something as being alive. Nothing lives without DNA. Every single living cell, without exception, possesses DNA at some point in its life. Built into the makeup of this molecule are the instructions for constructing and maintaining cells.

The structure of DNA

Watson and Crick's model of DNA showed that a DNA molecule is similar to a twisted ladder; it's a double helix. It is made up of two long chains, each composed of many smaller subunits called **nucleotides.**

Each nucleotide is made up of a sugar joined to a phosphate and one of four **bases.** These bases are adenine (A), cytosine (C), guanine (G), and thymine (T). The sides of the ladder contain alternating units of phosphate and sugar. The rungs are made up of bases. One of the first clues to figuring out the structure of DNA was that the amount of cytosine present in the molecule always equalled the amount of guanine, and thymine always equalled adenine.

DNA duplication

A cell duplicates its **chromosomes,** and therefore its DNA, before it divides. How is this done? DNA is not a single molecule but a double one. The double chain means that DNA has one characteristic that is essential to the continuation of life. DNA can reproduce itself.

Rather than thinking of DNA as a twisted ladder, think of it as a twisted zipper, with the pairs of bases as the interlocking teeth of the zipper. The bases on the DNA chains always pair up in the same way. Adenine is always paired with thymine, and guanine always pairs with cytosine.

When DNA is duplicated, the first thing that happens is that the two strands of DNA separate from one other. The zipper is unfastened. **Enzymes** bring more nucleotides to the two unzipped DNA strands. New nucleotides link up with the exposed bases on the unzipped strands, A to T and G to C. The new nucleotides then join together to form new strands attached to the original strands. The result is two double strands of DNA, rather than one.

The ability of DNA to replicate itself in this way is essential to cell division and reproduction. The **genetic material** has continued to be copied, from generation to generation, in an unbroken thread running back through time. This DNA connection goes back beyond dinosaurs and trilobites to the very first cells.

Rosalind Franklin

In 1951, the scientist Rosalind Franklin used a technique called X-ray diffraction to obtain images of the DNA molecule. She directed a beam of X-rays at a crystal of DNA and captured the image on film. She then used this to figure out the positions of the atoms in the molecule. Her results strongly suggested that the molecule had a helical structure. In 1953, Watson and Crick used Franklin's pictures to figure out that DNA spirals into a double helix. Watson and Crick were awarded the Nobel Prize in 1962 for this work. Sadly, Franklin did not share in this because she died of cancer at age 37, and the Nobel Prize is not given posthumously.

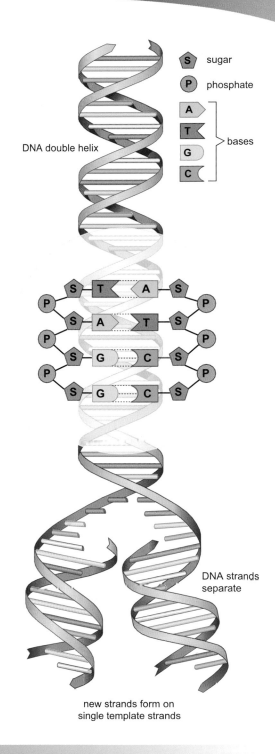

DNA double helix

S sugar
P phosphate
A
T } bases
G
C

DNA strands separate

new strands form on single template strands

Shown here is the structure of the DNA molecule and how it replicates itself.

The Genetic Code

The structure of **DNA** contains a four-letter code of instructions. But why? The instructions are recipes for making **proteins,** the substances that control the day-to-day activities of cells.

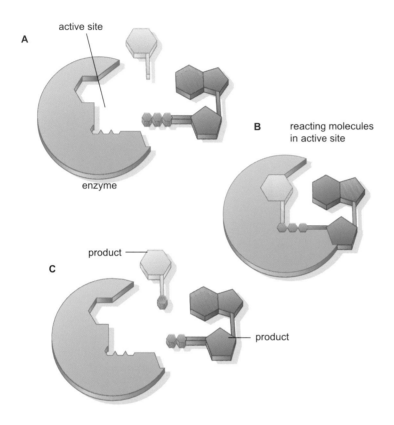

active site

A

B reacting molecules in active site

enzyme

product

C

product

The cell's managers

Proteins are key substances in every **organism.** Many of the structures in the body are made of proteins. For example, muscle fibers are proteins, and proteins are part of every cell membrane.

Other kinds of proteins are even more essential to life. Proteins called **enzymes** are natural catalysts. Catalysts are substances that control the rate of chemical reactions in a cell. There are thousands of such reactions happening at any one time, as molecules are broken down to release energy or built to form the molecules that make up the living cell.

Enzymes increase the speed of many cell reactions. Others would not happen at all without enzymes. The shape of the enzyme molecule has a space called the active site (A), where the substances involved in the reaction fit exactly, like a key in a lock (B). After the reaction, the new products are released from the active site (C).

Without enzymes there would be no life. A healthy cell needs to have all the right enzymes to perform essential tasks, such as getting energy, moving, growing, and repairing itself. Every reaction that takes place in a cell has its own specific enzyme. Because the enzymes control the reactions, they effectively control the cell.

Proteins and DNA

Proteins are complex substances. Protein molecules are long chains built from smaller molecules called **amino acids.** There are about twenty different amino acids used in making proteins. Each type of protein has a specific recipe, or a fixed order of amino acids along its length. This recipe has to be followed exactly, every time, or the protein will not be able to do its job.

We can now see why DNA is so essential to the cell. It is the instruction manual that holds the combinations for all the proteins that an organism makes. The protein-making instructions are held in the sequence of the **bases** along the length of the DNA molecule.

The genetic alphabet

There are four bases in DNA: adenine (A), cytosine (C), guanine (G), and thymine (T). The four bases are the alphabet of the genetic code. There are about twenty different amino acids that make proteins, but only four bases in DNA. So how can four bases code for twenty amino acids? The answer is that each amino acid is coded not by one base, but a group of three. These groups are called base triplets. The triplet AGC, for example, is the code for an amino acid called serine.

Now we can begin to see how DNA is used in the cell. The sequence of bases along the DNA molecule is a series of triplets. Each triplet codes for an amino acid, and a sequence of triplets codes for a protein. But there is still a very important question to answer: How does this code of bases on the DNA get translated into actual protein molecules?

The Human Genome Project is a huge research program in which scientists are working to map the complete genetic code for the human genome (all 23 **chromosomes**). This scientist is looking at a tiny portion of the sequence.

Duplicate codes

Can you figure out how many ways there are to arrange four letters in groups of three? The answer is 64. This means there are 64 possible ways to arrange the four DNA bases in groups of three. The genetic code contains 64 triplet codes. Because there are only twenty amino acids, there are 44 spare triplet codes. Scientists have discovered that some amino acids are coded for by more than one triplet. Some have two codes, some have four, and some have six. There are also three triplets that act as stop signs, marking the end of a protein sequence. There is also a triplet that acts as a start signal, indicating the beginning of a stretch of code.

From DNA to RNA to Protein

The basic unit of inheritance, the gene, is actually a length of **DNA.** Now that we have learned more about the structure of DNA and the genetic code, it is possible to see that each gene is a section of DNA that codes for a **protein** (or part of a protein). There are thousands of genes on each of the cell's **chromosomes.** So the cell's DNA is not just an instruction manual full of protein combinations, it's a vast library, containing codes for thousands of different protein sequences.

A cell's DNA never leaves the **nucleus.** One reason for this is that the DNA molecules are too big. Each chromosome in the nucleus is a single DNA molecule. Although chromosomes vary a great deal in size, they are all very large, containing millions of **base** pairs. However, experiments show that proteins are made outside the nucleus, in the cell **cytoplasm.** So there must be a mechanism for copying parts of the DNA code and carrying them out of the nucleus.

Reading the code

During **interphase,** when a cell is not dividing, there is a great deal of activity going on inside the cell nucleus. In many locations, sections of the double-stranded DNA are unraveled to allow a process called **transcription** to occur. During transcription, sections of DNA (individual genes) are copied. Just as when DNA copies itself, free **nucleotides** line up and join to the DNA strand by pairing with the bases on the DNA. In this case, however, a fairly short stretch of DNA is copied, and the molecule formed is not DNA but a close relative called **RNA,** or ribonucleic acid.

This diagram shows how proteins are made. As **messenger RNA** passes through a ribosome, transfer RNA molecules bring amino acids to add on to the growing protein chain.

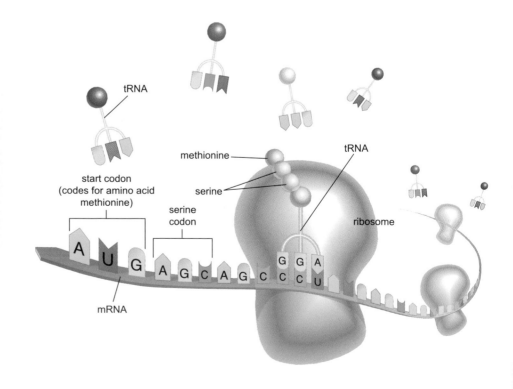

tRNA

tRNA

methionine

serine

start codon
(codes for amino acid
methionine)

serine
codon

ribosome

A U G A G C A G C G G A
                      C C U

mRNA

There are several different types of RNA. One, called messenger RNA (mRNA) carries a copy of the code for a particular protein chain. Messenger RNA is similar to DNA, but has only one strand, rather than two. Also, one of the bases in RNA is different from that in DNA. Instead of thymine (T), RNA has uracil (U).

Messenger RNA earned its name because it carries the code for a protein chain out of the nucleus and into the cytoplasm. In the cytoplasm, that code is read by tiny beadlike structures called **ribosomes.**

Assembling a protein

The making of a protein in the cytoplasm is very similar to a production line in a factory. Ribosomes attach themselves to an mRNA molecule at the start code for a particular protein. **Amino acids** are brought to the ribosomes by another type of RNA called **transfer RNA (tRNA).** Transfer RNA molecules are like forklift trucks. They bring the parts of the protein chain to the assembly line ready to be put together. Each tRNA molecule has one end that carries one particular kind of amino acid, and another end that joins up with the code for that amino acid on the mRNA molecule.

For example, say that a triplet on the mRNA is AGC, which codes for the amino acid serine. A tRNA molecule carrying the amino acid serine

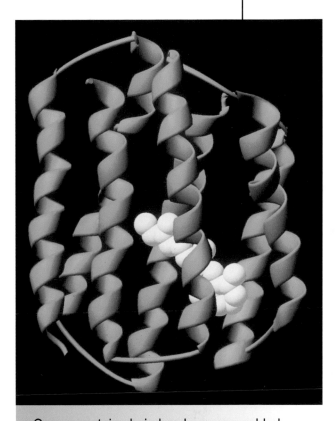

Once a protein chain has been assembled, it twists and folds to form a complex three-dimensional shape. This computer graphic shows the protein rhodopsin, one of the light-sensitive pigments in the eye. The blue is the protein chain, while the yellow is a light-sensitive compound called retinal.

attaches itself to an mRNA molecule. The ribosome then moves along the mRNA to the next triplet. If this triplet is also AGC, a second tRNA carrying a serine molecule will attach to the mRNA. The two amino acids are now joined together by an **enzyme.** The first tRNA then drops off, leaving the two serine molecules attached to the second tRNA. The ribosome now moves along to the third triplet on the mRNA, and the process continues. As the mRNA code is read, triplet by triplet, a long chain of amino acids is built.

It is possible for more than one ribosome to move along the mRNA at the same time as the cell busily assembles proteins. This is the genetic code at work.

Mutations

DNA replication is an incredibly accurate process. The DNA in a human cell contains about three billion **bases.** Cells divide time after time, and each time these three billion bases are copied correctly. However, very occasionally a mistake is made. The altered genes that result are called **mutations.** Usually cells that carry mutated genes die because they cannot function properly.

If a mistake happens as a normal cell is dividing, it will not have a great effect. The cell that has the mutation will probably die, and even if it survives, it is only one abnormal cell among millions of healthy ones.

The situation is different if the mutation happens during the formation of a **gamete.** If the mutated cell survives, it may go on to join with another gamete and form a new individual. The mutation may then be passed on to the next generation. A mutation that is passed on can cause serious problems. In humans, for example, a blood problem called sickle cell disease is the result of a mutation. People who have this disease have large numbers of abnormal, sickle-shaped blood cells that do not carry oxygen efficiently. Other mutations may not have a significant effect. Many people are color blind, for instance. This condition was originally produced by a mutation.

On rare occasions, a mutation can be beneficial and give an **organism** an advantage over its competitors. An example of this is when a mutation in a disease-causing bacterium gives it resistance to a drug. This mutation is beneficial to the bacterium, but not to humans with the disease.

Chromosome mutations and gene mutations

A mutation can affect an individual gene or an entire **chromosome.** A chromosome mutation involves a major alteration to one or more chromosomes, perhaps even a change to the number of chromosomes present. People who have the mental and physical disorder known as Down's syndrome were born with an extra copy of one chromosome.

This is the characteristic sickle or crescent shape of the red blood cells in a person with sickle cell disease. Magnification approx. x 5,200.

A gene mutation is a result of chemical changes in the cell's DNA. When a mutation occurs, the order of DNA bases that make up a gene is changed. This change can be very small, but it can have serious results. Sickle cell disease, for example, affects hemoglobin, the **protein** in red blood cells that carries oxygen. It is caused by changes in a single base pair in the genes that code for hemoglobin. This change results in the alteration of one **amino acid** in the hemoglobin, causing the hemoglobin molecule to be distorted.

A rare event

Most mutations are the result of a chance error that occurs when DNA is copied. Such an error is a rare event. A mutation occurs perhaps once in every 100,000 times that a DNA molecule replicates. However, in fast-reproducing organisms such as bacteria, some of which can divide every twenty minutes in ideal conditions, there are more chances for mutations to occur. If a few drug-resistant individuals appear in a population of bacteria, there will soon be a whole drug-resistant strain as the weaker individuals are killed and the resistant individuals are left to reproduce alone.

The likelihood of mutations occurring is increased by exposure to such things as ultraviolet light, X-rays, and some chemicals. Agents that cause mutations are called mutagens. There is no way to predict the type of mutation that will be caused by a mutagen.

Mutations and evolution

Most mutations that cause a visible change to an organism are harmful. However, on rare occasions a mutation improves an organism's chances of survival. If this individual then reproduces, it might pass this benefit on to its offspring. The offspring will also have improved chances of survival and, gradually, over many generations the mutation will be found in most members of the **species.** This kind of mutation is one of the processes involved in evolution, the way in which organisms gradually adapt and change over time.

This mutant fruit fly (*Drosophila melanogaster*) has four wings instead of the usual two. Fruit flies are widely used in genetic research because the DNA in their salivary glands is replicated many times to form giant chromosomes.

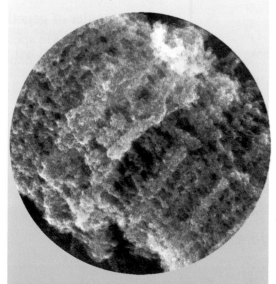

The patterns of bands in these giant fruit fly chromosomes can be used to identify different parts of the chromosome.

Phenotype and Genotype

Genes alone don't influence appearance. Genes interact with one another and with the environment to produce the outward form of the **organism** that we see. The actual appearance of an organism is called its **phenotype.** Underlying this phenotype are the genes that influence this appearance, called the **genotype.**

These prize-winning dairy cows have been bred for maximum milk production.

Homozygous and heterozygous

When the two **alleles** for a particular characteristic are the same, the organism is **homozygous** for that characteristic. If both alleles are dominant, the organism is described as homozygous dominant. If both are recessive, it is homozygous recessive.

If, on the other hand, the two alleles are different, the organism is **heterozygous.** If one member of the heterozygous pair is dominant over the other, the organism's phenotype will be the same as if it were homozygous dominant.

Breeding true

In his pea plant experiments, Mendel was careful to make sure that his plants always showed the characteristic he was interested in, generation after generation. When all the offspring have the same characteristics as its parents, this is called breeding true for that characteristic. Homozygous plants and animals will always breed true.

Farmers are particularly interested in making sure that the animals they breed are as good, if not better, than previous generations. If, for example, cows give a good milk yield, the farmer will want their calves to give a good yield, too. If a high-yield cow (MM) mates with a bull that also carries these alleles, all the offspring will be MM as well. However heterozygous animals (Mm) do not breed true. Of course, this is a very simplified example, as there are likely to be several genes that control milk production, plus environmental factors such as how the cattle are fed. In a simple case where a characteristic is controlled by one pair of alleles, what happens when we breed two heterozygous organisms together? We can find out by looking at another of Mendel's experiments.

Second-generation peas

After his experiment breeding purple-flowered peas with white-flowered peas, Mendel did a follow-up experiment. He bred together the purple-flowered offspring from the first experiment.

In modern terms, these plants were all heterozygous for flower color (Ff). If we think about what happens when the plants breed together, we can get some idea of what the results of this experiment might be.

Each of the pea plants would produce **gametes** with one flower color gene. Half of these gametes will carry the F allele, and half will have the f allele. When the plants breed, there are four possible results. The offspring could receive a dominant allele from both parents (FF) or a recessive allele from both parents (ff). They could get an F allele from the mother and an f allele from the father (Ff) or vice versa (fF).

Of these four possible genotypes, three will produce purple flowers, but one (ff, the homozygous recessive) would produce white flowers. For every four offspring, three should have purple flowers and one should have white flowers. Mendel's experiment did not produce such clear-cut results, but over many repeated experiments on thousands of pea plants, he did get close to a three to one ratio .

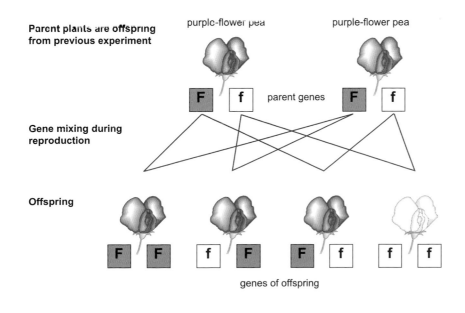

Parent plants are offspring from previous experiment

purple-flower pea   purple-flower pea

F   f   parent genes   F   f

Gene mixing during reproduction

Offspring

F   F   f   F   F   f   f   f

genes of offspring

This illustration shows a second experiment in which Mendel crossed the heterozygous purple-flowered offspring from the first experiment.

Is It a Boy or a Girl?

What is it that determines whether you are a male or a female? Your gender depends on a particular pair of **chromosomes** called the sex chromosomes. Females have a pair of identical sex chromosomes, called the X chromosomes. In males, the sex chromosomes are different. One is an X chromosome, as in females, but the other is a shorter Y chromosome. This means that females are XX and males are XY. When it comes to the sex chromosomes, for once the **genotype** has a strong bearing on the **phenotype.**

When males and females form **gametes (sperm** and **eggs),** the sex chromosomes are separated during **meiosis,** along with all the other chromosomes. In the female, every egg produced has one X chromosome. However, in the male 50 percent of the sperm will have an X chromosome and 50 percent a Y chromosome. If a Y sperm fertilizes an X egg, the result is XY, a boy. If an X sperm fertilizes the egg, the result is XX, a girl. Because there is a 50 percent chance of either event happening, the population will be about half males and half females.

Sex linkage

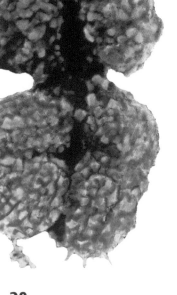

The sex chromosomes are different from all other chromosome pairs because the Y chromosome is shorter than the X chromosome. In fact, the Y chromosome has very few genes at all. So in a male, most genes on the X chromosome do not have a matching partner. There is only one **allele** for each gene.

This lack of a second allele explains why there are some inherited conditions that are much more common in males than in females. In humans, color blindness and the more serious condition of hemophilia, a disorder of the blood-clotting mechanism, are examples of this sex linkage.

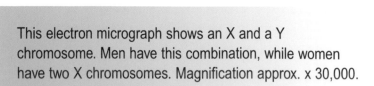

This electron micrograph shows an X and a Y chromosome. Men have this combination, while women have two X chromosomes. Magnification approx. x 30,000.

Queen Victoria of England was a carrier of hemophilia. Her son, Prince Leopold (second from the right), had hemophilia, and several of her daughters and granddaughters carried the disease.

The allele that causes red-green color blindness is recessive and is found on the X chromosome. If a woman has this allele, it will have no effect if she has a dominant allele on her other X chromosome. The dominant allele will mask the recessive one. She will only be color blind if she has two recessive alleles. If a man carries the recessive allele, the situation is different. There is no corresponding allele on the Y chromosome, so there can be no dominant allele to mask the effects of the recessive one. As a result, men are much more likely to be color blind than women. The same applies to the gene for hemophilia.

A woman can have one allele for color blindness or hemophilia and yet show no signs of either condition. A woman who has the allele for a disorder but shows no symptoms of it is called a carrier. In the case of color blindness, this is not so serious. However, hemophilia is a serious disease, so it is important to know if a woman is a carrier. If a man married a woman who was a hemophilia carrier, their children would inherit either a normal X chromosome or a Y chromosome from their father and a normal X chromosome or a hemophilia-carrying X chromosome from their mother. On average there would be a 50 percent chance that any son born would be a hemophiliac and a 50 percent chance that any daughter would be a carrier of the disease.

Different sex genes

Male and female **organisms** are not all distinguished genetically in the same way as in mammals. In some **species,** for example, females have the Y chromosome. In honey bees, males (the drones) are **haploid.** They only have a single set of chromosomes rather than **homologous pairs.** The males develop from unfertilized eggs, while the female eggs are fertilized by the queen bee.

Variety Is the Spice of Life

There is a huge range of dog breeds, from tiny dogs such as the Pekinese to tall, long-legged hunting dogs such as the Irish wolfhound. But all dogs belong to the same **species.** Any dog can breed with another. The differences in dog **phenotypes** that we see are called variations in the species. We can see variation in other species, including humans. Some species simply have a greater range of variation.

Variations that can be passed from one generation to the next are called inherited characteristics. They are **genetic** in nature. Other variations are due to environmental factors. One person might sit eating in front of the TV all the time and not exercise much. Another might play sports, eat well, and develop muscles. Muscle development is an acquired characteristic. Genes alone cannot give you big muscles. And you cannot pass on big muscles to your offspring. Many variations we see in **organisms** are a result of genetic and environmental factors working together.

Gene combinations

How does all this variety come about? **Mutations,** mistakes made when **DNA** is copied, are one way. An example is the appearance of cats with curly-haired coats, called the rex variety. This is thought to have arisen as a result of a mutation.

The main reason why one individual varies from another is because each one possesses a different combination of genes. This is partly due to how **gametes** are formed. Pure chance determines which member of a **homologous pair** of **chromosomes** will end up in a particular gamete. This is called free assortment.

These dogs show the wide variation in color, size, and other characteristics possible among different breeds of dogs.

40

Genes that are on the same chromosome are not separate in the same way as genes on different chromosomes. However, in one stage of the process of **meiosis,** there is a chance for genes from the homologous chromosomes to mix and match.

Crossing over

In the first stage of meiosis I (**prophase** I), pairs of homologous chromosomes join together to form double chromosomes, each with two **chromatids.** At this point, a process called crossing over takes place. In crossing over, each chromatid can break somewhere along its length. When this happens, the broken pieces can join up again as before. Or the pieces may swap over with the same pieces in their homologous partners. In other words, the chromatids swap genes. When the chromatids eventually separate to form gametes, there are new combinations of genes on each chromatid. This process of crossing over is unique to meiosis.

This process can be made clearer with an example. Before crossing over, the joined homologous chromosomes consist of four chromatids: two copies of each homologous chromosome. During crossing over, a section (B) from one chromatid swaps with a section (b) on its partner. On the other pair of chromatids, a section (C) on one chromatid swaps with a section (c) on its partner. So after crossing over all four chromatids are different.

The process of crossing over creates a much greater range of variation among gametes. Which particular male and female gametes join to form a fertilized **egg,** and therefore which set of genes is inherited from each parent, is random. This genetic variation is reflected in the physical variety of individuals.

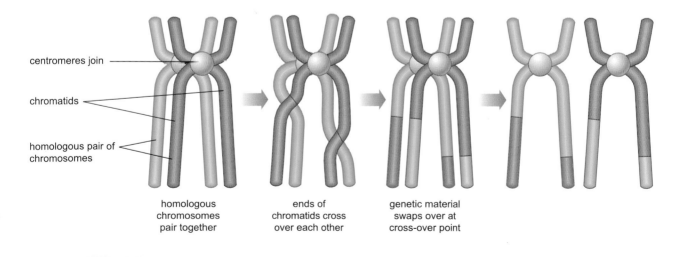

centromeres join

chromatids

homologous pair of chromosomes

homologous chromosomes pair together

ends of chromatids cross over each other

genetic material swaps over at cross-over point

This illustration shows crossing over during meiosis.

Genes and the Environment

Genes are not the only thing that determines an **organism's** appearance. The place where an organism lives and the events in its life also have a great effect. The influence of genes and the influence of the environment are closely linked. The two interact with each other and are difficult to separate.

Effects of the environment

We have already looked at one effect of environment, the effect that exercise has on muscle development. A more important influence on many **species** is how much food is available. Animals that have a limited food supply, especially when they are young, are generally smaller than animals that have plenty of food. However, these characteristics are not passed on to the animal's or plant's offspring because they are not coded into the genes.

The environment can have other effects on how an organism develops. In some cases, environment determines the sex of an animal. The sex of a crocodile or a turtle, for instance, depends on the temperature of the **egg** while it is incubating. Even more remarkably, fish called blue-headed wrasse have only one male in each school of fish. If the male dies, one of the females in the school will change sex and become a male.

Genes and evolution

Variation among the individuals of a species is a result of **mutations** and different combinations of genes. This variation has an important purpose. A group of organisms of the same species living in the same area is called a population. Some combinations of genes give rise to characteristics that give an advantage to an individual organism in the struggle for survival. Those individuals best equipped to survive in their environment are the ones that are most likely to have offspring.

Charles Darwin drew the beaks of different Galapagos finches. These birds were early colonizers of the islands. Several species arose, each having adapted to eating a different type of food.

Over millions of years, genetic changes in the ancestors of these dolphins and other marine mammals have enabled them to adapt to life in the seas.

This filtering out of the less well-adapted organisms is known as natural selection. Genes with a positive effect are selected in favor of genes with a less beneficial effect. Natural selection makes it possible for a population to adapt and survive if its environment changes, or if it has to move to a new environment. As an example, look at the way that modern giraffes are thought to have evolved their long necks.

The ancestors of today's giraffes had necks no longer than other similar animals. But food must have become scarcer for some reason. Perhaps the population of giraffe ancestors grew too large, or maybe other types of animals began to compete with them for food Another possibility is that the climate began to change, and the giraffes' food did not grow so well in the new conditions.

Genetic variation within the giraffe population produced some individuals that had slightly longer necks. These taller giraffes were able to reach food that the other giraffes could not. So more of them survived and produced offspring.

Natural selection and genetic variation have been the driving force behind the evolution of living things since the first cells appeared. Together, they have produced the enormous diversity of life we see today.

microtubule tiny, hollow protein tube that has the ability to shorten or lengthen.

mitochondrion (plural mitochondria) structure within eukaryote cells that is responsible for energy release

mitosis type of nuclear division in which the number of chromosomes stays the same

mutation change in a cell's genes, a chromosome's structure, or in the number of chromosomes in a cell

nucleotides small subunits from which DNA and **RNA** are made. Nucleotides consist of a sugar, a phosphate, and a base.

nucleus structure where the cell's genetic material is stored

organelle membrane-bound structure within a cell that is responsible for specific jobs within that cell

organism any type of living thing

phenotype outward appearance of an organism and the characteristics that result from its genetic makeup

photosynthesis process by which green plants and some other organisms use the energy of sunlight to make food from carbon dioxide and water

pollination part of sexual reproduction in flowering plants in which pollen grains (containing male gametes) from a flower are carried to the female reproductive part of the same or another flower

prokaryote cells that do not have a nucleus. Bacteria are the only prokaryotes.

prophase first stage of cellular and nuclear division in which the chromosomes first become visible through a microscope

protein substance that makes up many cell structures and controls their reactions

protistan large, loose grouping of organisms, many of them single-celled

ribosome structure within eukaryote cells that is responsible for protein synthesis

RNA (ribonucleic acid) substance related to DNA that carries out a number of important tasks in protein synthesis

species organisms of the same species can breed together and produce fertile offspring

sperm male sex cell

spindle structure made from microtubules that attaches to the chromosomes and pulls the two copies of each chromosome apart during cellular and nuclear division

stigma part of the female reproductive structure in flowering plants to which pollen grains attach

telomere structure on the end of a chromosome that protects it from damage during cell division

telophase fourth and final phase of cellular and nuclear division in which new nuclei are formed

transcription production of a single-stranded RNA molecule from a section of DNA

transfer RNA (tRNA) type of RNA that carries amino acids to the ribosomes and is involved in attaching them to make proteins

vesicle small, round cavity in the cell surrounded by a membrane

zygote fertilized egg cell

Further Reading

Morgan, Sally. *Science at the Edge: Genetic Modification of Food.*
    Chicago: Heinemann Library, 2002.

Microlife Series: four books about the world of microorganisms
Snedden, Robert. *A World of Microorganisms.* Chicago: Heinemann
    Library, 2001.
Snedden, Robert. *Scientists and Discoveries.* Chicago: Heinemann Library,
    2001.
Snedden, Robert. *The Benefits of Bacteria.* Chicago: Heinemann Library, 2001.
Snedden, Robert. *Fighting Infectious Disease.* Chicago: Heinemann Library, 2001.

Wallace, Holly. *Life Processes: Cells and Systems.* Chicago: Heinemann Library,
    2002.

Index